METALS AND ALLOYS

Number four in the *Materials We Use* series

D1808503

METALS AND ALLOYS

R. W. Thomas B.Sc
Head of the Upper School
The Norton Knatchbull School, Ashford, Kent

Illustrated by R. F. Summers

WHEATON
A Member of the Pergamon Group

OXFORD · NEW YORK · TORONTO · SYDNEY

PERGAMON PRESS LTD, HEADINGTON HILL HALL, OXFORD,
OX3 0BW
PERGAMON PRESS INC., MAXWELL HOUSE, FAIRVIEW
PARK, ELMSFORD, NEW YORK 10523
PERGAMON OF CANADA LTD, P.O. BOX 9600 DON MILLS,
ONTARIO, M3C 2T9
PERGAMON PRESS (AUST.) PTY LTD, 19A BOUNDARY
STREET, RUSHCUTTERS BAY, N.S.W. 2011, AUSTRALIA

Printed in Great Britain by A. Wheaton & Co., Exeter

ISBN 0 08 017877 4

CONTENTS

The object of this series is to describe the basic materials used in everyday life and the even larger number of materials which can be made from them. It is obviously impossible to describe materials without bringing some chemical and other scientific principles into the story and I have tried to explain the reasons for the various processes described. The reader will find, however, that he needs to know little formal science and some of the principles referred to may have been explained in other books of the series. These books are just as suitable for pupils in the lower forms of secondary schools as for adult readers who want to know something of the materials around them even though they may have forgotten much of their school science.

In this book we turn our attention from the organic carbon-containing compounds, which have formed the basis of the three earlier titles in the series, to the metals and alloys. In *The Chemical Giants* it was described how plastics have replaced metals in many fields, but there will always be a vast number of uses for which metals will be the first choice. As with all technological processes there must always be competition in the supply of the best material for a particular purpose and so it is not surprising that there is constant change.

My thanks are due to all those individuals and organisations who have helped me with information and advice and in particular the following who have supplied illustrations:
Aluminium Federation Figures 26, 28.
British Aluminium Company Ltd Figures 27, 29.
British Steel Corporation Figures 12, 14, 16.
Consolidated Gold Fields Ltd Figure 35.
Copper Development Association Figure 33.
Imperial Metal Industries Ltd Figure 39.
Imperial Smelting Corporation Ltd Figure 20.
International Nickel Ltd Figures 1, 3, 5, 7, 18, 21, 30, 37.
Shell Chemicals U.K. Ltd Figure 17.
Steetley Company Ltd Figure 36.
U.K. Atomic Energy Authority Figure 23.
Zinc Development Association Ltd Figures 19, 24.

R. W. Thomas
Ashford, Kent

1 WHAT IS A METAL?

Most of you would think that you could easily say whether a given piece of solid was a metal or not. How do you decide? You might say that the sample was shiny (but so are diamonds which are certainly not pieces of metal) or that it conducts electricity (as do the black rods in the middle of torch batteries but these are made of graphite, another form of carbon like diamond). You might add that the piece of solid could be made into wires or beaten into sheets and that when you dropped it it made a ringing noise—but so does a piece of steel.

"Of course," you will say. "Everyone knows that steel is a metal". To the scientist it is not however. All types of steel contain iron, together with varying amounts of other elements, usually carbon. A substance produced by dissolving one or more elements in a molten metal and then cooling the mixture is known as an alloy. Consequently, steel is an alloy, and when the scientist talks about a metal he refers only to an element which has metallic properties. Therefore, iron is a metal whereas steel is not.

We have learnt in other books in this series that there are just over a hundred known elements. More than three-quarters of these are metals but even so this means that there are only eighty or so metals. When you think just how many combinations of metals (and even of metals with non-metals) there are you will realise that there is an almost unlimited number of possible alloys which could be made.

Fig. 1 A huge travelling crane capable of lifting up to 125 tonnes holds these steel sheets by means of electromagnets. Look carefully at the picture to see how many different uses of metals you find in it.

1

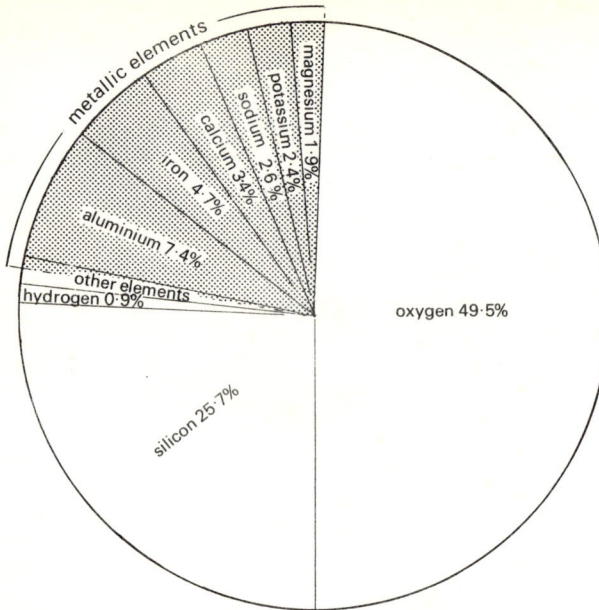

metallic elements

magnesium 1·9%
potassium 2·4%
sodium 2·6%
calcium 3·4%
iron 4·7%
aluminium 7·4%
other elements
hydrogen 0·9%

oxygen 49·5%

silicon 25·7%

Fig. 2 The relative abundance of elements in the earth's crust. The figures shown are only approximate and are based on the masses of the elements.

Assuming that you accept all this we still have not explained how we can decide whether an element is a metal or not. For most purposes the five properties which have been used for many years to decide this worked quite satisfactorily. For example, most metals

1 conduct heat and electricity well
2 are malleable i.e. they can be beaten into sheets
3 are sonorous i.e. they ring when struck
4 usually have high melting and boiling points
5 show certain chemical differences to non-metals

If you consider some of the elements, you would not have much difficulty in deciding that, say, copper and iron were metals whereas oxygen and nitrogen, the main gases in the air, were not.

There are a few elements which may offer more doubts but it is unlikely that you will come across them often.

Although it is sometimes convenient to be able to decide whether a solid is a metal, alloy or non-metal by means of a brief examination, today the scientist concerned with metals (called a metallurgist) is far more concerned with the structure of these substances. A metal is now usually taken as being a crystalline substance containing positively charged ions (which you will remember are atoms or groups of atoms carrying an electric charge) embedded in a sort of 'sea' of negative electrons. The charges on the ions are neutralised electrically by the opposite electric charge carried by the electrons so the lump of metal is not electrically charged. Moreover, the crystals are usually so tiny that they do not usually look like crystals, even under a microscope. However, if you look at a piece of galvanised iron (such as is used for making metal dustbins) you can often see the crystal patterns in the zinc coating.

You may well have seen an experiment in which a piece of zinc dipped in a solution of a lead salt causes the lead metal to come out of solution (or be displaced) as crystals.

Fig. 3 A highly magnified view of a chromium/nickel alloy. The smaller particles contain mainly chromium and only a small amount of nickel whereas the remainder is rich in nickel.

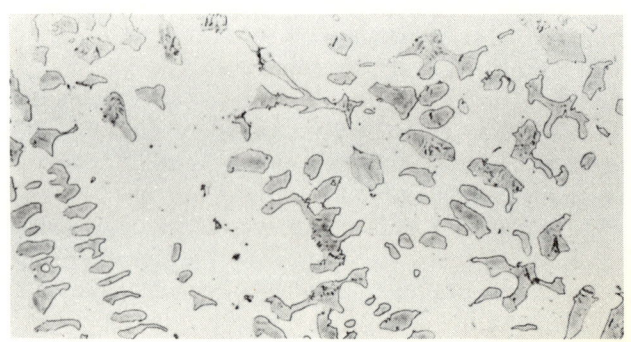

Suggested activities

1 Make a list of as many objects around you as you can which appear to be made of metal. In each case try to find out what metal or alloy they contain.

2 A number of characteristic properties of metals and non-metals have been listed in this chapter. Can you think of any others which might help you to distinguish between metals and non-metals.

3 You have read that a scientist who works mainly with metals is called a metallurgist. With the aid of books in the library, make a list of as many industries as you can who might employ metallurgists.

2 WHERE DO METALS COME FROM?

A few metals such as gold, silver, platinum and occasionally copper can be found in nature uncombined with other elements. In most cases, however, chemical reactions have to be carried out to obtain the metal and, even in the case of the precious metals mentioned, it is rarely easy to obtain the pure substance.

Minerals from which metals can be obtained in sufficient quantities to be worthwhile are known as ores. These are often found mixed with large quantities of unwanted rocky material and so the first stage in most processes for producing metals is the concentration of the ore. Even when the ore has been obtained in a reasonably pure state it may still not be in a form from which the metal can be obtained directly. In this case the ore may have to be changed to another chemical compound from which the metal can then be extracted.

The next stage is the actual production or extraction of the metal itself. Although this is perhaps the most dramatic stage of the whole operation it is not necessarily the most difficult.

Finally the metal may have to be purified. In some extraction processes the metal is obtained in a reasonably pure state and may be suitable for most purposes without further treatment. In other cases even tiny amounts of impurities make the metal useless. As we shall see on page 33, this is so when germanium is being produced for use in transistors.

You should now realise that a mineral may have to undergo a long and expensive series of processes before the metal can be obtained. This means that not every mineral containing a particular metal can be used to produce that metal. One of the most important factors in deciding whether it is worthwhile to extract a metal from a mineral is the cost of obtaining that metal. There is obviously no point in setting up a complex plant for obtaining a metal from a newly discovered mineral source if the cost of extracting a tonne of that metal is twice the price at which other companies are selling it.

For example, to obtain a metal such as iron which is used in vast quantities and so needs to be relatively cheap, only ores with a fairly high proportion of iron in them would be used. Certainly no ore with less than 20% of iron in it would normally be considered and even then only if all other conditions were right. On the other hand it is worthwhile producing gold from a rock even though a hundred tonnes of mineral may have to be treated to obtain one kilogramme of the metal (that is only one-thousandth of one per cent).

It sometimes happens that the discovery of new or improved methods of extracting metals means that a mineral, once regarded as useless, can become quite valuable. In the same way that an article which your grandfather may have bought for few pence can now be worth many pounds, so mineral deposits which were once ignored may now become valuable.

At one time, price was almost the only factor considered when deciding how to obtain a metal.

Today we are more conscious of the fact that our mineral resources will not last forever and that in the past valuable materials have been squandered. In the future greater care must be taken to conserve our natural resources. As the steel industry has shown for many years this can be done profitably by reusing scrap iron but other industries have often not bothered.

In the next part of this chapter we shall take a closer look at each of these stages of metal production and then, in the remainder of the book, some of the more important metals will be discussed in greater detail.

Concentration of ores

The ores must first be removed from the Earth. Often the deposits are found close to the surface as in the large ironstone areas in the Midlands. Here the minerals can be dug out using huge dragline excavators as shown in Figure 4. In recent years a great deal more care has been given to avoid spoiling the countryside;

Fig. 4 A huge dragline excavator being used to remove surface material so leaving the mineral exposed. The size of the 'bucket' can be judged by the bulldozer on the ground below.

Fig. 5 Mines are not always the most attractive of places. This is a large open pit mine in the Sudbury district of Ontario in Canada.

after a large area has been mined, it is restored to agricultural land and, within a few years, there is no sign of any mining activity at all. In some cases the layers of ore deposits may be thick enough to justify going deeper and deeper forming an enormous pit.

More often the required mineral is found thousands of metres below the surface in which case a system of shafts and tunnels has to be built. The further the mines are cut into the rock, the longer the tunnels become, until eventually they may become so complex that, as in the

Fig. 6 Setting the fuses for an underground blast.

Fig. 7 In these large rotating drums, lumps of minerals are crushed by spinning them with hard steel rods.

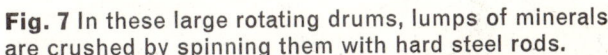

Zambian copper mines, they may extend for hundreds of kilometres. The transport of men and materials, ventilation of the tunnels and ensuring that these do not collapse then becomes a feat of engineering and organisation in itself.

Having obtained the ore, it usually needs further concentration. There are several ways of doing this depending on the properties of both the minerals and of the waste material in which the ore is found (known as gangue). Most of these involve crushing the rock to a fine powder.

The most commonly used method of separating the substances is called flotation. The finely powdered material is run into tanks of water to which has been added a chemical, similar to the detergents described in another book of this series.

7

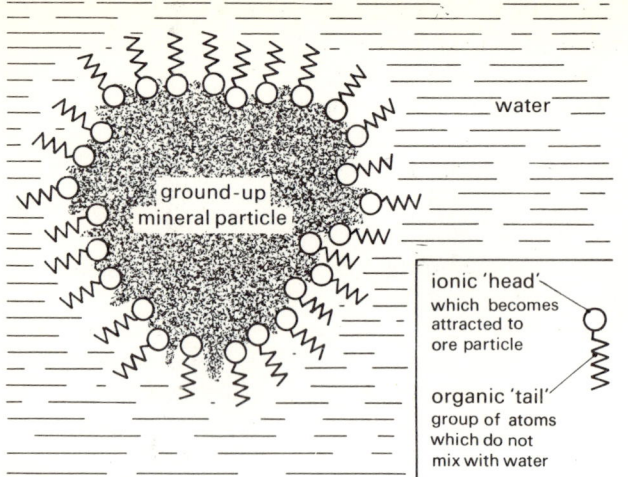

water

ground-up mineral particle

ionic 'head' which becomes attracted to ore particle

organic 'tail' group of atoms which do not mix with water

Fig. 8 The diagram shows how the small grains of mineral are prevented from mixing with the water so leaving them free to collect in the froth.

Fig. 9 A large batch of flotation cells in which the waste material is skimmed off in the froth.

These substances contain molecules, one end of which consists of a chain of carbon atoms and the other a group of atoms which is attracted to the particles of ore. Each particle is, in effect, surrounded by 'organic' groups of atoms which do not easily mix with the water (See Figure 8). The result is much as if the particles were coated with grease in that the water does not wet them. The substance added does not stick to the gangue particles so well and these sink to the bottom of the tank. If air is blown through the water to which has been added a foaming agent, a froth is produced and the mineral particles remain in this froth. This can then be skimmed from the surface taking the ore particles with it.

The iron oxide mineral magnetite has, as its name suggests, magnetic properties and so can be separated from the waste rock by means of powerful magnets.

Further treatment of the ore

Most minerals require chemical processing before passing on to the extraction stage. There are two main reasons for this. One is that the mineral may contain large quantities of impurities which have not been removed in the concentration stage either because they are chemically combined with the required metal compound or because their properties are so similar to those of the metal compound that the physical methods described above cannot be used. For example the common aluminium ore bauxite consists of aluminium oxide mixed with iron oxide which, as we shall see later, would interfere with the production of aluminium. Fortunately the aluminium oxide is more acidic than the iron oxide and so the former can be dissolved in a solution of sodium hydroxide which is strongly alkaline. The iron oxide remains undissolved and can be filtered off; the aluminium oxide can easily be obtained from the remaining solution.

Another common reason for the need to use chemical treatment is where it is impossible to extract the metal directly from the compound in the mineral. For example, many metals are found in nature combined with sulphur as metal sulphides These are not easily reduced to the metals but most of them readily change to metal oxides when heated (or roasted) in a stream of air. These oxides can then be converted to the metals as described in the next section.

Extraction of the metal

There are three main ways in which a metal can be extracted from its ore.

Oxides of the less reactive metals (see Figure 10) can usually be converted to the metal by heating them with carbon (in the form of coke) or with carbon monoxide (usually produced from carbon during the process as is described on page 14 in connection with the production of iron). The removal of oxygen from a compound is an example of reduction and a substance which brings this about is called a reducing agent. Carbon and carbon monoxide both readily react with oxygen (to form carbon dioxide gas) and so are acting as reducing agents. There are many other reducing agents but usually these are too expensive for use in large-scale production processes. For example hydrogen is an excellent reducing agent which suffers from high cost but it is used for extracting a few metals where expense is not so important.

In some cases the metal chlorides can be reduced to the metals by removing the chlorine with more reactive metals. The big snag with this method is that the more active metals such as sodium, calcium and magnesium are themselves quite expensive and so this method can also only be used for metals such as titanium which can be sold at high prices. Moreover, chlorides of these metals are not often found in nature and so have to be produced from other minerals which adds further to the cost.

Methods such as those described above in which the metallic compound is heated with a reducing agent are called smelting processes. Simple smelting techniques have been used for thousands of years. Compound of the more active metals could only be reduced, as explained above, by using even more reactive metals which are

Fig. 10 The electrochemical series; in general the most reactive metals are found near the top of the list.

Electrochemical series

Potassium	*Nickel*
Calcium	*Tin*
Sodium	*Lead*
Magnesium	*Copper*
Aluminium	*Silver*
Zinc	*Mercury*
Iron	*Gold*

9

extremely costly. Only a hundred years ago, aluminium was produced in this way and was so expensive that it was regarded as a precious metal suitable for making jewellery.

Early in the Nineteenth Century it had been shown that some metals could be made by passing electricity through solutions of salts (in the same way that you may have produced copper from copper sulphate solution). The very reactive metals could not be obtained in this way since the hydrogen in the water was given off more easily than the metal in solution. In 1800 Sir Humphrey Davy showed that sodium could be made by passing electricity through sodium hydroxide which was kept at a high enough temperature to melt it. It was however a long time before electricity became cheap enough to make electrical methods possible on a large scale. Even thirty years ago it was only those countries which could produce cheap hydroelectric power (usually from mountain streams) which used these processes. The development of a relatively cheap electrical method for making aluminium in 1885 caused an immediate drop in price and within a short time saucepans of this metal became common. The decomposition of chemical compounds by electricity is known as electrolysis.

Nowadays those metals which are produced by electrical methods are usually obtained by the electrolysis of their molten chlorides, since chlorides of the more active metals (unlike those of the less active ones) are widely found in nature.

A third technique for extracting metals which, until recently has only been used to a small extent, but which now seems likely to bring about a revolution in metal production is known as solvent extraction. For a long time gold has been obtained by dissolving the metal from the crushed ore by using a solution of sodium cyanide and then displacing the gold from the solution by adding zinc dust. The sodium cyanide solution was therefore acting as a solvent for the gold which you may remember is present in only tiny quantities in the ore. The solution produced contained a much greater amount of gold.

The discovery of new solvents has enabled this technique to be used for obtaining quite concentrated solutions of metals from deposits which a few years ago would have been considered useless. It is already being widely used for extracting copper.

You can now see that obtaining metals is rarely a simple business and often it is very complicated indeed. Thus among the factors which decide the value of a metal are not only how common are its ores but how easily and cheaply the metal can be obtained from these ores. In the next few chapters we shall be taking a closer look at a number of metals but you should remember that since there are over eighty metals we have space to examine only the more common and the more interesting ones in this book.

Suggested activities

1 By means of reference books and atlases try to find out some of the areas where the minerals of metals mentioned in this book are found. You may be able to persuade your Geography master to produce an outline map of the World and then you could mark these areas on this map

2 Simple smelting techniques have been used for thousands of years. What metals or alloys were known and produced so long ago. How were they obtained? How did they affect the lives of the people at the time? You may find it useful to consult History books for this.

3 In some parts of the country you may find evidence of ancient metal-working. For example maps showing the Weald of Kent and Sussex have many references (usually as place names) to the early iron industry which was so important in this area. Can you find signs of such activity in your area? A study such as this is known as industrial archaeology and this is now becoming a popular hobby for many.

4. Go to your local museum to see if they have samples of minerals. If so, you may be able to make a series of coloured drawings showing typical crystal shapes.

3 IRON AND STEEL

There is a good reason why we should start our detailed look at some of the metals with a study of iron. The quantity of iron and steel produced throughout the World far exceeds that of any other metal and, moreover, iron has been worked in this and many other countries for thousands of years. The industry has naturally undergone many changes since the Iron Age but the same basic principle of iron production, namely the extraction of iron from oxide ores by the removal of oxygen still applies. The modern iron industry in the United Kingdom really dates from the 1730's when Abraham Darby designed a blast furnace using coke (or more accurately the carbon monoxide produced from it) as the main means of reducing the ore. Before that charcoal, obtained from wood, had been used as the main fuel.

The most important sources of iron are the oxide ores, haematite, limonite and magnetite, which contain between 50% and 75% of the metal. The ore found in this country consists mainly of iron carbonate and is of low quality containing little more than 20% of iron. Thus Great Britain relies to a very large extent on imported iron. This is not an unusual situation. Many of the World's biggest iron and steel producing countries use ore brought in from outside. For example, Japan, which is a bigger iron producer than the U.K., has little iron ore of its own and even the mighty American steel industry has to import some ore. Modern iron and steel works are therefore nearly all situated close to deep water harbours so that big iron ore carriers can unload their cargoes close to the works. Another important source of the metal is scrap iron. Not only is the price of iron obtained from this lower, but it does not have to be bought from outside the country. We have already seen how important it is for the World to conserve its natural resources and the recycling of iron in this way helps a great deal. Unfortunately, since iron and steel consumption is higher now than it was perhaps twenty years ago when the present scrap was 'new' iron, there is not enough to go round.

Production of iron

As we have seen, the essential features of the iron producing process have changed little in 200 years but the modern furnace is far bigger and more efficient than those built perhaps fifty years ago.

These huge blast furnaces are the most obvious feature of any iron and steel works as they tower over thirty metres in height. They consist of vertical steel towers lined with fire-resistant bricks. The shape of the furnace is shown in Figure 11 but as Figure 12 reveals this is concealed by a complex array of girders and pipes around the furnace. The charge fed to the top of the furnace consists of coke, iron ore and limestone. At one time these were added as separate loads but today it is more common to use sinter or pellets which are lumps containing ore, coke and often limestone fused together. To make sinter, ground-up ore is mixed with powdered coke and limestone and heated in a stream of air.

Fig. 11. A diagram of a typical modern blast furnace.

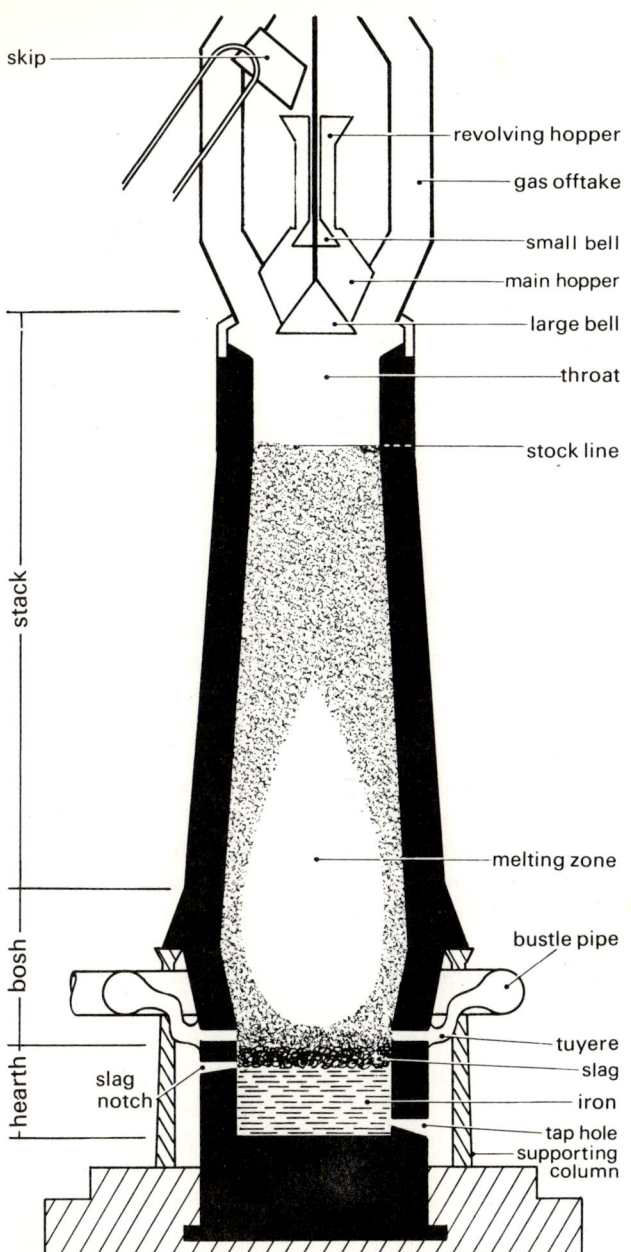

skip

revolving hopper

gas offtake

small bell

main hopper

large bell

throat

stock line

stack

bosh

hearth

slag notch

melting zone

bustle pipe

tuyere

slag

iron

tap hole

supporting column

As some of the coke burns the mixture becomes fused together. Pellets are made by rolling powdered ore and limestone with a little water in a large rotating drum. The lumps are mixed with powdered coal and heated to harden the pellets.

The coke used in the charge has to be of a particularly hard type so that it does not become crushed to a powder which might clog the furnace. It is made by heating coal in coke ovens (without any air present or the coke would burn away). The gases given off are certainly not wasted and can be burnt to produce heat for making sinter or for raising steam to drive pumps or produce electricity.

Fig. 12 The No. 2 blast furnace at Llanwern Works of the British Steel Corporation, showing the vast array of tubes and girders which complicate the basically simple outline. The large tube in the top centre of the picture carries the hot waste gases from the furnace. The three towers on the right are a set of hot stoves.

The other substance needed is air. Although this is obviously plentiful a single furnace may need 4,000 tonnes of air a day and this has to be pumped in by huge steam turbines so it is certainly not free. The air is blown in (providing the 'blast' of the blast furnace) near the bottom of the furnace through a series of pipes or tuyères which are fed from a large bustle pipe encircling the furnace. The air is heated before entering by means of hot stoves. These contain large amounts of fire-brick which is pre-heated by the hot exhaust gases coming from the top of each blast furnace. When a set of stoves becomes sufficiently hot, air is passed through them on its way to the furnace. This air is then heated until the stoves cool down. Three stoves are generally used for each furnace so that two are being heated up while the third is pre-heating the air. One of the most common ways of cutting down the cost of an industrial process is to reduce waste of heat and you will find throughout this series examples of ways in which 'waste' heat is used for other purposes. A look at your home gas or electricity bill will reveal how expensive heat is these days and the use of waste heat can save a great deal of money in a large-scale process.

The simplest way of describing what happens in the blast furnace is to start at the point where the hot air comes in contact with the coke. This burns vigorously giving out a lot of heat (as in a coke boiler) and producing carbon dioxide gas. This is the hottest point of the furnace. As the carbon dioxide passes up the tower it passes over white hot coke which takes some of the oxygen from the carbon monoxide. Both coke and carbon dioxide are thus changed to carbon monoxide gas. Further up the furnace where the temperature is a little lower this carbon monoxide picks up oxygen from the iron oxide ore so converting it to iron and in so doing changes back once more to carbon dioxide. As this iron gradually moves down to the hotter part of the furnace it changes from a spongy solid to a liquid which trickles down through the burning coke to collect on the hearth. A common impurity in iron ore (as in many other minerals) is silica or silicon dioxide which you will know best as the main ingredient of sand. This would still be solid even at the hottest part of the furnace and, since blast furnaces are kept going more or less continuously, it would be difficult to remove. This is why limestone is included in the original charge as this substance reacts with the silica to form calcium silicate which melts to a liquid. This then trickles to the bottom of the furnace. Fortunately this liquid does not mix with the molten iron but forms a separate layer on top rather like oil and water. The substance formed is known as a slag.

This process is continuous but slow; it may take a specimen of iron ore eight to sixteen hours from the time of entering the furnace to the point where it reaches the hearth as molten iron. Every few hours this molten iron can be tapped from the furnace through holes which are usually kept closed by clay plugs. A few feet above these is another set of holes through which the slag can be tapped. Some of this slag contains compounds of phosphorus when it is known as basic slag and can be used as a fertilizer but mostly it is used for roadmaking or as railway ballast. The days when slag was merely dumped as a waste product are now fortunately nearly past.

The iron leaving the furnace contains up to 4% of carbon (from the coke) and is very brittle. It is usually referred to as pig iron or cast iron. Some of it is run into moulds to form blocks of iron (called pigs) but much of it is transferred as molten iron direct to the steel works which nowadays are nearly always on the same site or nearby.

Huge quantities of hot waste gases leave the top of the furnace by means of the large pipes

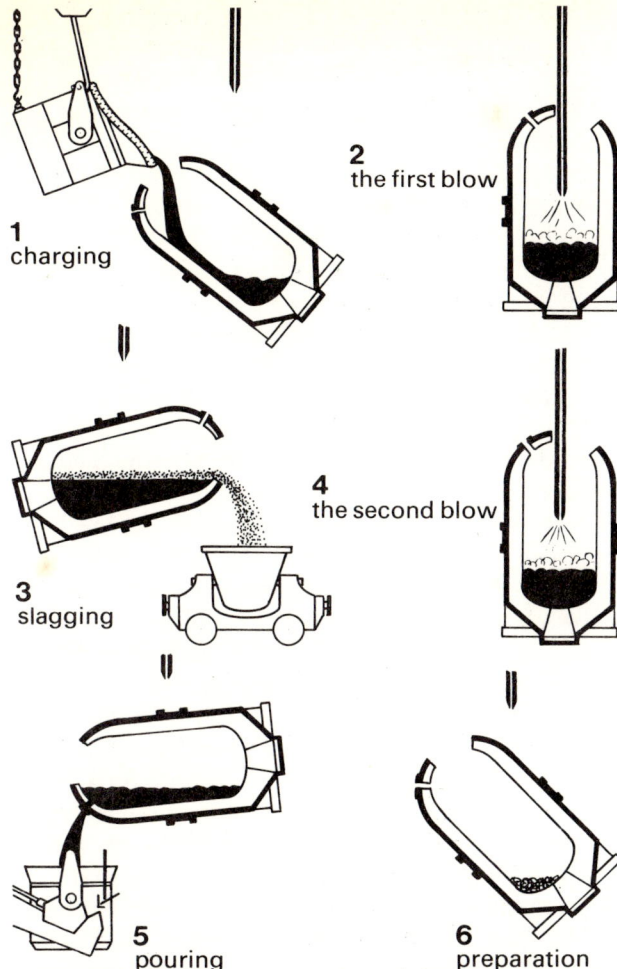

1 charging

2 the first blow

3 slagging

4 the second blow

5 pouring

6 preparation

Fig. 13 These diagrams show the operation of the LD or BOS process.

In modern blast furnaces the air supply often has extra oxygen added to it and fuel oil is injected; this means that less coke, which is now very expensive, is needed. Today a furnace may produce as much as four times as much iron in one day as a typical furnace of only twenty-five years ago.

We have seen that the pig iron from the furnace is hard but brittle and so is unsuitable for many purposes. Consequently a high proportion is converted to steel by removing most of the carbon; there are many types of steel but the most common are the carbon steels which contain up to 1% of carbon.

The usual methods of making steel today are those in which the surplus carbon is changed to carbon dioxide gas by means of a jet of pure oxygen. In the LD process or BOS (basic oxygen steel) process, as it is usually known in this country, over a hundred tonnes of molten iron is run into a large steel tub or converter which has a lining of refractory bricks. A high speed jet of oxygen is then blown onto the surface by means of a tube or lance. There is an awe-inspiring gush of flame from the mouth of the converter as the carbon in the iron is oxidized. A hundred tonnes of iron can give up perhaps four tonnes of carbon in less than twenty minutes. Since the process is so fast the industry needs very rapid and efficient ways of analysing the steel to discover how much carbon is left in it and to ensure the process can be stopped at the correct point. The molten steel is then poured out by tilting the converter. Other impurities in the iron are removed by adding lime (calcium oxide) to form a slag similar to that from the blast furnace during the process. The various stages are shown in Figure 13. Unfortunately some of the iron is changed to iron oxide which is carried off as a reddish smoke and expensive equipment has to be installed to prevent this polluting the countryside.

called downcomers. These gases are mostly nitrogen (from the air blast) together with carbon monoxide and carbon dioxide (from the burning coke) and are mixed with a fine dust of iron ore and coke which can be removed and sintered. The heat from these gases is used as described above to preheat the air blown into the furnace.

Fig. 14 An LD converter in operation. In this country it is usually referred to as the BOS process.

Until recently most of the steel in this country was made by the Open Hearth process. In this as much as four hundred tonnes of steel could be made by heating a mixture of cast iron, limestone (to form a slag), iron oxide ore (to provide the oxygen to remove carbon) and varying amounts of iron and steel scrap. Often a jet of oxygen is used to speed up the process but even

Fig. 15 A simplified diagram of an electric arc furnace. Heat is produced by the arc of electricity between the carbon electrodes shown.

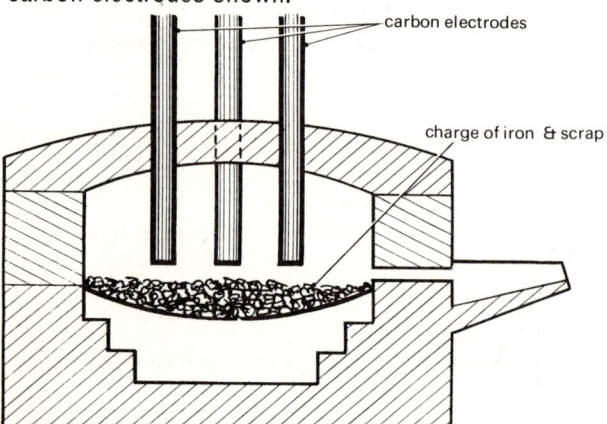

carbon electrodes

charge of iron & scrap

so it is far slower than the oxygen converter processes described above and Open Hearth furnaces are now being rapidly replaced.

There are other ways of making steel but space only allows us to mention the Electric Arc furnace. In these furnaces heat is produced by passing electricity between large carbon rods dipping into the materials. This method is particularly suitable for treating scrap iron and steel and a plant of this type has recently been set up in Kent to produce steel from scrap; in the past this would have had to be carried long distances to steel works which could have used it.

The molten steel is then poured into large moulds made of cast iron to form slabs known as ingots which may weigh many tonnes. These ingots can then be pressed while still red hot between rotating rollers, much as a domestic wringer presses clothes. By passing the hot metal through a series of presses the ingot can be change into long strips, sheets or girders according to requirements. A modern continuous rolling mill has these sets of rollers arranged at intervals along a track several hundred metres long, the red hot slabs being moved along by resting on a large number of rotating rollers. In some modern steel works the molten steel is converted directly into strips without being moulded into ingots first. This is known as the continuous casting of steel.

We have already noted that most of the World's steel production is centred on the manufacture of carbon steels. There are many types of these which cover a large range of possible uses. In general, mild steels (which contain very small amounts of carbon) are relatively soft but easily worked whereas increasing the proportion of carbon makes the steel harder but rather more brittle. For example, a high carbon steel with 1 to $1\frac{1}{2}\%$ carbon in it might be used for tools or rails whereas a mild steel would be better for nails or wire.

Fig. 16 An electric arc furnace in operation.

Although the total production of steels containing other elements (known as alloy steels) is tiny compared to the quantities of carbon steels manufactured, their value is high since many of them are extremely expensive because of the cost of the elements added. Again there is a vast range of possibilities but the table on p. 18 shows a number of examples together with some of their uses.

One of the main snags of iron and steel is their tendency to rust. This occurs when iron comes in contact with oxygen and water so causing it to change to a form of iron oxide. The process occurs particularly quickly when there is something dissolved in the water which helps it to conduct electricity better since rusting is basically an electrical process. This is why cars rust so badly in a few weeks of wintry weather when salt has been put on the road to prevent icing.

Fig. 17 Every year millions of pounds are lost through the rusting of iron and steel. The mass of this swing bridge must remain virtually unchanged since its balance is so critical. The steel was prevented from rusting by coating it with a protective metal. After erection, it was cleaned and touched up. A further coat of stainless steel was then applied and this was protected further by the use of a special polymer.

Fig. 18 The tyres of this 200 tonne scraper train used in road-making have been protected by chains of nickel steel which has been specially hardened on the surface. The ones shown had already been in use for over 3500 working hours.

Rusting can therefore be prevented by putting a suitable coating over the iron. This could be of oil in the case of moving parts (which only provides a temporary answer), paint or another metal (such as tin, zinc on galvanised iron, or chromium) in other cases. Alloy steels which contain a lot of chromium do not rust easily and are known as stainless steels. If a strip of zinc or magnesium, which are more reactive metals than iron, is bolted to the iron it will 'dissolve' more readily and so prevent the iron from rusting. Such metals are known as sacrifice metals.

Much money is spent every year on protecting iron and steel from rusting or on replacing the corroded parts. Indeed it has been suggested that in the United Kingdom alone tens of millions of pounds are spent every year for this purpose.

A small book like this only enables us to take a brief look at what is one of the World's most important industries. If you want to find out more about it, *Iron and Steel* by R. W. Thomas (Pergamon Press) goes into the subject in rather more detail.

Suggested activities

1 From which countries does the United Kingdom import most of its iron ore?
2 A number of different types of steel have been referred to above. Try to find out what elements are present in as many types of steel as possible. You may also be able to find how their properties vary (how hard? can they stand up to high temperatures? etc.) and for what purposes they are used.
3 In this book we have only been able to refer briefly to the methods of working steel (i.e. turning it into wires, girders, sheets, etc.). You will probably find many books in your library which describe these processes. Write a short description of the more important ones using diagrams to help you if you can.
4 What ways are there of preventing iron or steel from rusting? Try to find as many examples as possible in your house, garden or school and work out why a particular method has been chosen in each case. You might start by looking at dustbins, garden tools, cars etc.

ALLOY STEELS AND THEIR APPLICATIONS

MAIN ALLOYING ELEMENTS	TYPICAL APPLICATIONS
Manganese	*Railway points*
Tungsten	*Cutting tools*
Chromium	*Ball bearings, bright parts on cars, cutlery*
Nickel	*Drive shafts, measuring instruments*
Vanadium	*Tools and springs*
Cobalt	*Magnets*
Titanium	*Gas turbine blades, spacecraft and rockets*

Although iron and steel are used on a vast scale throughout the World, their greatest disadvantage is the ease with which they rust away unless protected. One of the methods mentioned earlier was galvanizing, in which the iron is coated with a layer of zinc, either by dipping, spraying or electroplating. Another process described was the use of sacrifice metals, of which zinc is one. Other uses of zinc include roofing sheet, the outer cases of torch batteries and in the making of many alloys.

As well as being an important metal it is a suitable one to discuss at this point in the book. It was explained earlier that a number of metals are obtained from their ores by removing the oxygen combined with them, for example by the action of carbon or (as in the case of iron) by carbon monoxide. Zinc is another example of such a metal. The chief sources of zinc are the minerals calamine (zinc carbonate) and zinc blende (mainly zinc sulphide) which is often found mixed with galena (lead sulphide). Before these can be converted to zinc they have to be changed into zinc oxide which is easily done by heating, or roasting, them in air. Mixed ores of zinc and lead presented problems until the discovery of one of the processes described below.

If the oxide formed is heated with carbon (in the form of coke) the oxygen is removed leaving zinc metal. At one time this was done by heating the mixture in a large number of small clay retorts but this is no longer done in this country since it was wasteful both of fuel and manpower.

Fig. 19 Iron sheet emerging from a galvanizing bath.

Today continuous processes are widely used and these fall into two main types, the vertical retort and blast furnace processes.

In the vertical retort process the powdered zinc oxide is mixed with ground coal and heated to form briquettes. These are then heated more

19

Fig. 20 Molten zinc is tapped from the furnace.

strongly driving off the gases from the coal leaving the briquettes porous. These are fed into the top of the retort while still hot. The vertical retorts are towers made of very hard wearing carborundum bricks. They are some seventeen metres high and last over three years on average. The retorts are heated by means of circulating burning fuel gases which heat the towers from the outside. The carbon reacts with the zinc oxide in the briquettes producing zinc (as a vapour at the temperature of the retorts) and carbon monoxide gas. The zinc is condensed outside the retort as a liquid and the carbon monoxide can be used to heat the retorts. The process is continuous in that a load of briquettes can be added to the top of the retort every hour or so while liquid zinc is run off from the condensers at regular intervals—there is no need to stop the heating of the retort.

Small amounts of another metal, cadmium, are usually mixed with the zinc and since the metals have different boiling points they can be separated by fractional distillation, a higher temperature version of the process used for separating the constituents of crude oil (as described in an earlier book of this series, *Petroleum and Natural Gas*). Since there may be only one part of cadmium to every two hundred parts of zinc, it is not a cheap metal. Its uses include the plating of radio and electrical equipment and in cadmium/nickel batteries. When mixed with some other metals it forms very low melting alloys. A use of one of these is to make the plug in automatic fire sprinkling systems; as the temperature rises as the result of a fire the plug melts so allowing water to pass through the pipes and to spray onto the fire from the sprinklers.

It has already been stated that zinc and lead are often found together in the same mineral and although they are both valuable metals they may be difficult to separate. A blast furnace process enables both metals to be extracted at the same time! In some respects the process resembles that described for the extraction of iron in that a sinter of oxides (made by roasting the original minerals) and coke is fed into the top of the furnace while a blast of air is forced in near the bottom. The

Fig. 21 The electric power for the Swingfire anti-tank missile, seen here leaving its launcher, is provided by a nickel-cadmium battery.

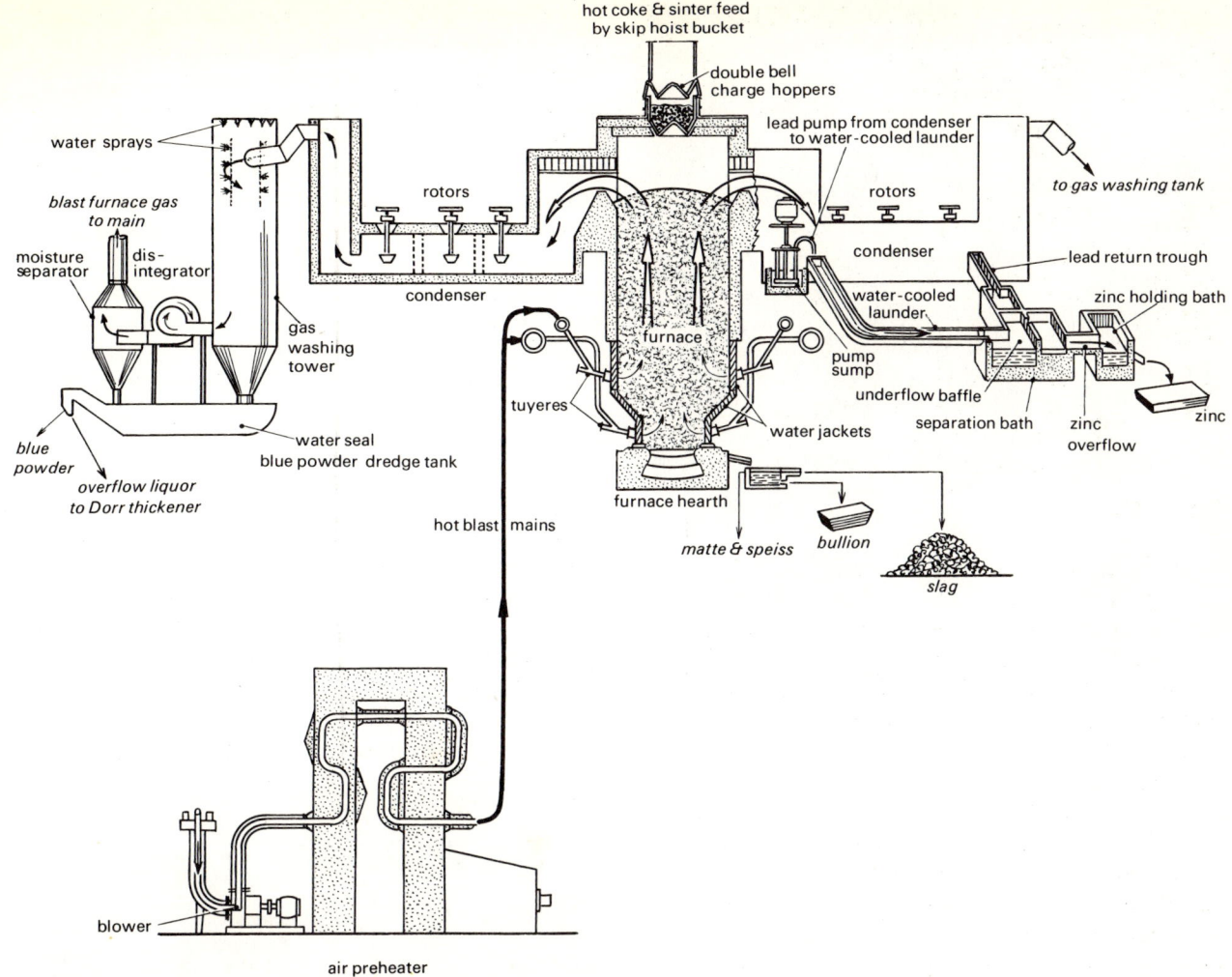

hot coke & sinter feed
by skip hoist bucket

double bell
charge hoppers

water sprays

blast furnace gas
to main

moisture
separator

dis-
integrator

rotors

lead pump from condenser
to water-cooled launder

rotors

to gas washing tank

condenser

lead return trough

condenser

gas
washing
tower

zinc holding bath

furnace

water-cooled
launder

blue
powder

pump
sump

zinc

underflow baffle

water seal
blue powder dredge tank

overflow liquor
to Dorr thickener

tuyeres

water jackets

separation bath

zinc
overflow

zinc

hot blast mains

furnace hearth

matte & speiss

bullion

slag

blower

air preheater

Fig. 22 A simplified diagram of a zinc/lead blast furnace.

difference between this and the retort process is in the way the metals are removed from the furnace. Very conveniently zinc boils at a much lower temperature than lead and so, at the temperature of the furnace, the zinc vapour passes out near the top whereas the lead remains as a liquid. Thus the metal can be tapped from the bottom of the furnace much as in the iron blast furnace. Consequently the two metals are separated in the furnace during the extraction stage but the overall process is a good deal more complex than this and great care is taken to avoid the pollution of the atmosphere by lead.

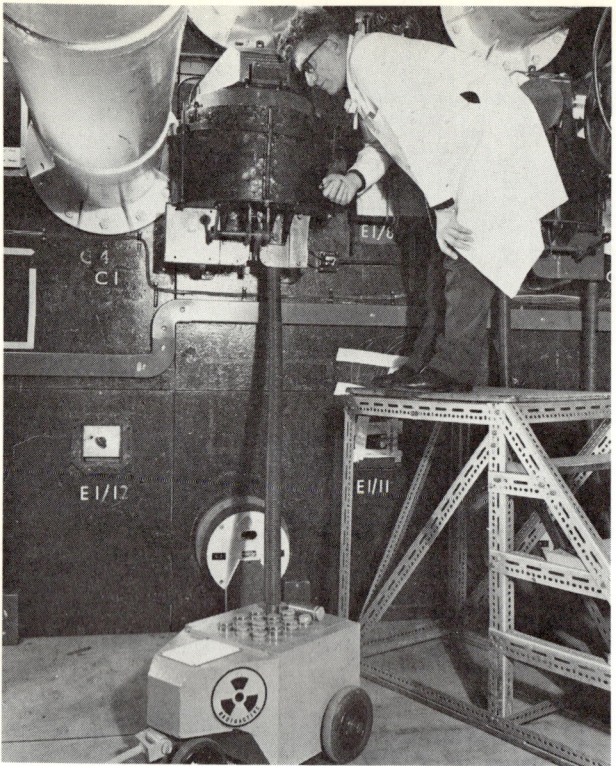

in low-melting alloys such as typemetal and solders. It is also used as protective cases for radioactive materials.

Yet another way of obtaining zinc is by passing electricity through a solution of its salts. For example, zinc ores can be dissolved in sulphuric acid and electricity passed through the zinc sulphate solution produced; zinc is produced at the negative electrode which is usually made of aluminium. This is an experiment which can easily be carried out in a school laboratory by passing low voltage electricity (NOT the mains voltage) through a solution of zinc sulphate using carbon rods as electrodes. The negative electrode or cathode soon becomes coated with a silvery-looking deposit of zinc metal. The process can use those ores which contain little zinc and so are not suitable for the furnace processes but, like all electrolytic processes, it is only suitable when large quantities of cheap electricity are available.

Fig. 23 Radioactive materials being removed from a reactor at Harwell into a lead container which offers protection from the radiations.

Fig. 24 Some idea of the huge scale of industrial electrolytic processes is given by this photograph of the cell room in a zinc plant. The sheets of aluminium which act as the cathode are seen being removed after being covered with zinc.

This process, therefore, produces useful quantities of lead which is more usually obtained from sulphide ores such as galena in suitable furnaces. Because of the poisonous nature of lead and many of its compounds some of the uses of these substances have declined as people have become more concerned with the harmful properties of the materials they use. Whereas at one time nearly all paints were based on lead compounds, now very few are. Nevertheless, lead is still an important metal and is used widely in building chemical plant (since few substances corrode it), for the electrodes in car batteries and

This is perhaps a good place to point out that sulphuric acid (as used above and for many other purposes) can be obtained as a valuable byproduct of zinc and lead production. When the sulphide ores are roasted in air to produce the oxides, sulphur dioxide gas is given off in large quantities and this can readily be made into sulphuric acid. This is yet another example of the need for the chemical industry to reduce waste to a minimum; in this case by making use of what is from the point of view of zinc production, really a waste product, but is in fact a valuable source of an important chemical. Unfortunately, the use of by-products in this way may not always be so simple!

Suggested activities

1 Why do you think that the large continuously working furnaces now used for producing zinc are so much more efficient and economic than the small retorts once used which had to be emptied and reloaded every few hours?

2 Electrolysis is used in one method of producing zinc. From books in this series and others in the library, make short descriptions of as many other large scale processes using electrolysis as you can.

3 In this book there are mentioned several ways of reducing waste in industrial processes. Make a list of them together with any others you can find and explain why they are so important.

5 ALUMINIUM

It was explained earlier that aluminium is too active a metal to be extracted by methods similar to those described for iron, zinc and lead, Thus, until 1886 when a process for obtaining aluminium by electrolysis was invented, large amounts of aluminium were unavailable. It was then regarded as a precious metal whose price was more than £5 a kilogramme; this was despite the fact that aluminium is present in clay and makes up about one-twelfth of the Earth's crust. Today aluminium is always extracted from the mineral bauxite which contains the oxide of the metal.

The production of aluminium by electrolysis is not simple since passing electricity through a solution of an aluminium salt does not cause the metal to be deposited on the negative electrode as is the case with zinc and copper; instead hydrogen is given off. To overcome this problem

Fig. 25 The materials used to produce one tonne of aluminium.

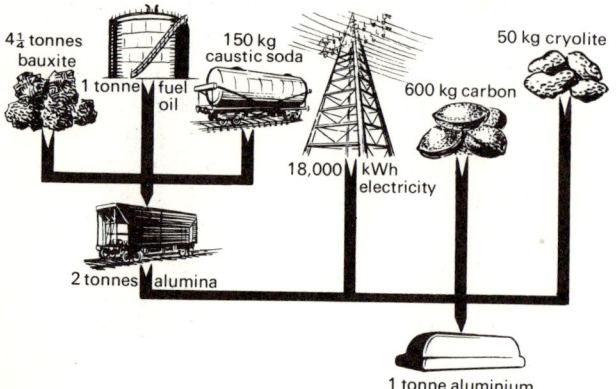

4¼ tonnes bauxite

1 tonne fuel oil

150 kg caustic soda

18,000 kWh electricity

600 kg carbon

50 kg cryolite

2 tonnes alumina

1 tonne aluminium

aluminium oxide (obtained from the bauxite) is dissolved in molten cryolite. This is another mineral containing aluminium but today it is rare and so is usually made artificially for this purpose.

As already pointed out, any electrolytic process needs cheap electricity, since the larger the amount of a substance produced, the greater the quantity of electricity required. No matter how efficient the process can be made this problem can never be overcome. Thus most aluminium plants have, in the past, been built where cheap hydroelectric power is available. It is now becoming more common however, to site them in other areas and to make use of other sources of fuel for producing the electricity, for example natural gas, coal or even nuclear power.

The first stage in the production of aluminium is to make the pure oxide from the bauxite (which also contains oxides of iron and titanium). This is done by treating the crushed bauxite with hot strongly alkaline sodium hydroxide solution. This dissolves the aluminium oxide, but not the other compounds present, so these can be filtered off. When the remaining solution is cooled a few crystals of aluminium oxide are added and this causes aluminium oxide to come out of solution as a solid. This contains water and so is heated to produce pure aluminium oxide as a white powder.

This aluminium oxide is now dissolved in the molten 'cryolite' in a shallow steel box lined with carbon. The carbon, which is a good conductor of electricity acts as the cathode (negative electrode) and the anodes (positive) consist of carbon blocks (see Figure 26) which are lowered into the liquid. When a heavy electrical current is passed, aluminium metal, which stays as a liquid at the temperature of the apparatus, collects at the bottom of the cell (or pot as it is known) and can be tapped off from time to time. At the anode, oxygen from the aluminium oxide is given off and causes the carbon blocks to burn away as carbon monoxide and dioxide gases. Thus the anodes are gradually used up and so occasionally have to be lowered further into the pot and eventually need to be replaced completely. Aluminium oxide is periodically added to the molten cryolite to replace that used up in the electrolysis.

Aluminium has several properties which differ from those of other common metals and it is these which make it so important. It is light (its density being only about one-third of that of iron) and is a good conductor of both heat and electricity. It is rather soft when pure, but readily forms alloys with other elements such as magnesium, silicon, copper or zinc to produce a stronger product; some aluminium alloys are as strong as steel.

Fig. 27 One of the 500 metre-long cell rooms at the British Aluminium smelter at Invergordon in Scotland. The carbon anodes ready for use in the cells can be seen on the right.

Aluminium readily reacts with the oxygen in the air to form its oxide. This also happens with iron, producing rust (see page 17) but, whereas iron rust flakes off the metal, the oxide layer on aluminium sticks tightly and so protects the metal. This oxide film, which can be made thicker by a process called anodising, can be dyed permanently so producing the attractive finishes seen, for example, on some saucepans and other kitchenware. Aluminium also polishes well and so is used for reflectors in headlamps and in mirrors.

Fig. 26 A simplified diagram of an electrolytic cell for producing aluminium.

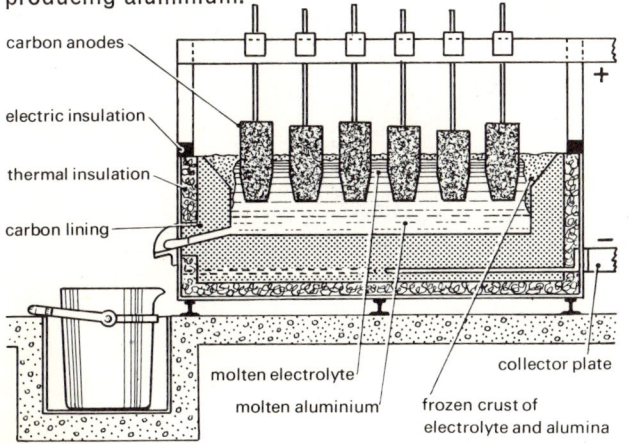

carbon anodes
electric insulation
thermal insulation
carbon lining
molten electrolyte
molten aluminium
frozen crust of electrolyte and alumina
collector plate
+
−

M.A.—C.

Fig. 28 An aluminium body for an experimental sports car showing its extreme lightness.

The lightness and high electrical conductivity already mentioned make it valuable for overhead cables since an aluminium wire can carry as much electricity as a copper conductor twice its weight. The lightness and strength of aluminium alloys are made use of in aircraft construction. Aluminium compounds are not poisonous and so the metal is widely used for cooking utensils and packaging (as foil). The metal is used widely in chemical plant and for road-transport tankers because of its lightness and resistance to a wide range of substances, many of which will corrode other metals. Aluminium is, however, readily attacked by hydrochloric acid and metal chlorides in solution and so sea water corrodes it rapidly.

When aluminium combines with oxygen in chemical reactions a great deal of heat is produced. Thus when aluminium powder is heated with iron oxides it combines with the oxygen in the latter and the heat produced is so enormous that it melts the iron formed; this is the basis of the Thermit process for welding iron.

Although aluminium is best known for the uses of the metal itself, some of its compounds are also of value. The oxide can be obtained in a very porous form known as activated alumina and this is useful for drying gases and as a catalyst for speeding up some chemical processes. Another form of the oxide, produced at high temperature, is very hard and is used as corundum for cutting tools and in polishing (as in emery paper). Incidentally this form of the oxide, combined with traces of other metal-salts which colour it, is found in nature as rubies and sapphires.

Aluminium chloride is used as a flux in soldering, keeping the joint clear of any oxides formed during the process. Alum or aluminium potassium

Fig. 29 Another picture showing the applications of many metals. The portable trackway supporting this crane at West Burton power station in Nottinghamshire is made of aluminium.

sulphate is used in dyeing and is also used for fire-proofing fabrics. Aluminium acetate is used in water-proofing cloth—if the material is soaked in a solution of this substance and treated with steam, the pores of the fabric become sealed with the aluminium hydroxide produced.

Suggested activities

1 A few of the applications of aluminium have been given above. Make a list of as many other uses of the metal (or its alloys) as you can, indicating where possible which of the properties of aluminium influenced its choice in each case.

2 It was stated above that aluminium oxide is the main constituent of rubies and sapphires. With the aid of an encyclopaedia or other reference books, make a list of a number of precious stones. State where you can the chemical elements or compounds each contains. What factors do you consider decide whether a substance becomes a precious stone?

3 One of the uses of aluminium oxide is as a catalyst. What is meant by this term? A number of other examples of catalysts have been given in this series. How many of these can you remember (or look up)?

6 COPPER

Like iron, copper is a metal which has been known and used for thousands of years. It is, however, only quite recently that it has been produced in the vast quantities used today and in 1911 the *Encyclopaedia Britannica* dismissed the huge copper ore deposits of what is now Zambia, with the words "There is also some copper . . ."

One of the main reasons why large-scale production was so slow in coming was that, although copper minerals are found widely and in large quantities, most of them contained only a tiny proportion of copper. Another snag was that these deposits were usually in what were then regarded as very remote areas.

An ore containing 3% or 4% of copper is considered high-grade; whereas you will recall that iron producers are rarely interested in ores containing less than 20% of the metal and even these are often ignored. Many of the copper ores worked today contain less than 1% of copper which means that more than 99% of what is mined may be waste material! Consequently much of the effort and expense in copper production goes into the refining of the ore before the actual extraction of the metal can start.

There are many different minerals containing copper and some have attractive colours. They fall into two main categories, those in which the copper is combined with sulphur often in company with other metals, and the oxidized ores in which the metal is present as the oxide or carbonate. Among the main producers of copper today are the USA, Zambia, USSR, Chile and Canada.

In some cases the presence of other metals in the ores is a nuisance, whereas in others it may prove to be a blessing. In the early days of the now famous Sudbury mines near Lake Huron in Canada it was found difficult to extract the copper from the deposits and a new metal, later identified as nickel, was found in the furnaces. At that time the problem was how to dispose of the nickel whereas today this metal is extremely valuable (usually for alloying with other elements as in cupro-nickel coinage metal) and more nickel than copper is produced at Sudbury. Incidentally, these Sudbury mines also yield many thousands of tonnes of precious metals, such as gold, platinum and silver.

Fig. 30 Tapping the matte from a nickel furnace.

After digging out the minerals, they are crushed and concentrated by flotation. The principles of this process were described on page 7. At the end of the process the powdered copper ore collects in the froth and so can be skimmed off. The copper sulphide may be mixed with iron and or nickel sulphides; it is then placed in large roofed tanks known as reverberatory furnaces which are heated by means of a mixture of powdered coal or oil and air blown in from above. The charge in the furnace gradually separates into two liquids, a molten mixture of copper, iron and possibly nickel sulphides at the bottom, covered by a layer of molten slag. When the slag has been drawn off, the molten matte as it is known can be poured out. In the Sudbury works there are, as we have seen, considerable quantities of nickel sulphide in the matte; the latter is treated with sodium sulphide which causes the matte to separate into two layers, the top one containing mainly nickel sulphide and the lower one mainly copper sulphide. The two portions can then be processed separately.

In some more recent plants the oxygen is blown into the furnace instead of the fuel/air mixture described above. The ore reacts so violently with the oxygen that enough heat is produced to keep the process going. In this way large amounts of coal or oil are saved; this process is known as flash smelting.

The molten matte is then fed into large tubs known as converters. Air can be forced through

Fig. 31 Converters for the production of nickel or copper.

the mixture and huge quantities of evil-smelling sulphur dioxide gas are given off; this can be made into sulphuric acid. The molten charge gradually changes to a mixture of copper with a slag containing mainly compounds of iron. The slag is first tapped and then the molten copper can be run off. The copper is now 98–99% pure but the small proportion of oxygen contained in it makes it unsuitable for many purposes and so the copper (known as blister copper because of the appearance of its surface) needs further refining. One way of removing the oxygen which is still used is to put green branches of hardwood trees into the molten metal.

The other method of purifying copper is by electrolysis. The blocks of blister copper are used as anodes which are dipped into large tanks containing copper sulphate solution. The cathodes consist of thin sheets of pure copper which fit between the anodes. The copper in the anode dissolves when electricity is passed through and pure copper is deposited on the cathodes. Thus these become thicker as the anodes gradually

Fig. 32 Stacks of copper 'wirebars' being loaded for transport at a copper mine in Zambia.

dissolve away. Impurities in the blister copper often include the precious metals mentioned above and these fall to the bottom of the tank from which they can be dredged as the so-called anode slime. When the cathodes reach a suitable size they are lifted from the tanks and replaced by a further set of thin plates.

The so-called oxidized ores are easier to treat and are usually heated in a blast furnace with coke. The oxygen in the ore reacts with this to form carbon monoxide gas so leaving molten copper which can be tapped as is iron in the blast furnace.

A process used for dealing with large quantities of ores containing only tiny proportions of copper compounds is leaching. This involves allowing large volumes of very dilute sulphuric acid to drain through big piles of crushed rock so producing a dilute solution of copper sulphate. Since there is so little copper in this solution it has been expensive to handle in the past but the new process of solvent extraction described on page 10 has now made it far cheaper.

In this method a relatively small amount of a suitable organic liquid (which does not mix with the solution) is brought into contact with the copper sulphate solution. The liquid takes up copper very readily so that a far more concentrated solution is formed and this is easily separated from the aqueous layer since they do not mix. The organic liquid is then treated once more with sulphuric acid so that the copper is again changed into copper sulphate but this time the solution is far more concentrated. If this solution is electrolysed, copper is deposited on the cathodes as described above. Solvent extraction is thus a most economic way of obtaining a workable solution of copper sulphate from ores containing so little metal that they would be impossible to process in any other way; the organic liquid can be used over and over again. As described earlier, this solvent extraction process has made it possible to extract copper (and other metals) from deposits which a few years ago would have been regarded as useless.

Perhaps the most important property of copper is its high electrical conductivity (only silver, which is far more expensive, being better) and thus the main uses of the metal include the manufacture of wires, cables and electrical and electronic equipment in general. It is, however, now finding aluminium a strong competitor in overhead wires where the lightness of aluminium

Fig. 33 For hundreds of years copper has been used as a roofing material. On prolonged contact with the air it takes a pleasant green colour. On this new poultry market at Smithfield in London, ten thousand sheets were used weighing thirty-five tonnes in all.

is an important advantage but, for most other conductors, copper leads the field in spite of its high cost. A large transformer for example may need over forty tonnes of copper in its windings.

On exposure to air, copper quickly becomes coated with a green deposit of a copper compound which protects the metal from further corrosion and gives it a pleasing appearance. Consequently it has been used for roofing public buildings for many hundreds of years and this use still continues today. Perhaps the most important use of copper in building is for making the pipes and tanks used in modern hot water systems.

For thousands of years, copper alloys such as bronze (copper and tin) have been known and copper alloys are still used widely. Perhaps the best known use of a copper alloy is in coins. 'Silver' coins (which have contained no silver for many years) contain 75% of copper in the cupronickel alloy used, whereas bronze coins contain 95% copper. Other examples of the uses of copper alloys are in car radiators, chemical, brewing and distillation plant, and for ships' propellers.

The compounds of copper also have their uses. Most important is the inclusion of copper sulphate and other compounds in sprays for the treatment of rotting wood and preventing plant disease and an oxide of copper in anti-fouling paints for protecting the hulls of ships.

Suggested activities

1 Why do you think that copper alloys are used so widely throughout the world for making coins? Why have silver coins been replaced by ones made of a copper/nickel alloy? Why is pure copper not used for coins?
2 Discuss the relative values of copper and aluminium (a) in roofing (b) in making overhead electrical cables (c) in making kettles.
3 If you were asked to build a plant to produce very pure copper from its ores what factors would you have to take into account when you decided where to erect it?

7 MORE METALS

In a book of this length it is obviously not possible to describe all the known metals in detail. There are, however, a considerable number of other metals which are important and about which you should certainly know something. Some of these are produced in large quantities whereas the total production of others, which may be no less important, may only amount to a few hundred tonnes a year. In this chapter we take a brief look at some of these other metals; what you have learned earlier will enable you to appreciate the methods by which they are obtained even though only an outline can be given here.

Chromium. The main chromium ore is chromite in which the metal is combined with both iron and oxygen. The metal is usually made by heating chromium oxide (obtained from the chromite) with aluminium. This combines with the oxygen, producing so much heat that the chromium metal left can be run off as a liquid; the principle is similar to that used in the Thermit process described on page 26.

The main uses of the metal are as protective coatings and in alloys such as stainless steel which is less prone to rusting than are ordinary steels. When steel is chromium plated it is usually first coated with copper and/or nickel to help the chromium to adhere to the steel and to improve the resistance to corrosion. Some of the more important uses of chromium compounds are in the tanning of hides, as solutions in chromium plating baths, and, since many chromium compounds are highly coloured, as pigments in paints and dyes.

Cobalt. This usually occurs in nature combined with arsenic and sulphur and is frequently mixed with nickel and copper compounds. Indeed much cobalt is obtained as a by-product in the production of copper and nickel from Canadian sources (see page 28). The pure metal can be obtained from the oxide by reducing it with carbon, hydrogen or, as in the case of chromium, with aluminium.

Some of its compounds are coloured and so are used in pigments. The main use of the metal itself is in alloys with a considerable range of different metals that are extremely hard and resistant. Cobalt steel is used in permanent magnets.

Germanium. Although germanium is regarded commercially as a metal its electrical conductivity regarded at room temperature is very low. At temperatures above 100°C however, it is a good electrical conductor. These unusual electrical properties cause germanium to be labelled as a semi-conductor and, when mixed with tiny controlled amounts of other elements, it is used in transistors.

Germanium is a rare metal and is usually obtained as a by-product of many industrial processes. For example, in Great Britain germanium is obtained as the oxide from flue dusts formed when Northumberland coal is burnt.

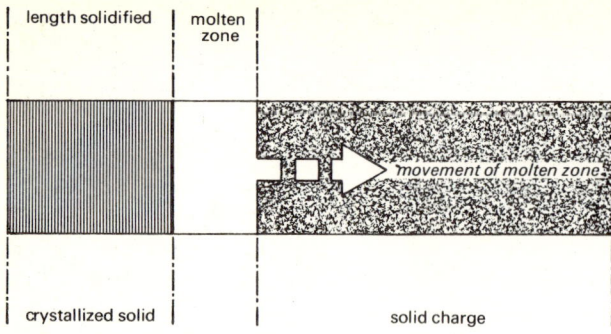

length solidified | molten zone

movement of molten zone

crystallized solid | solid charge

Fig. 34 A diagram representing the process of zone refining. The impurities tend to collect in the molten zone which gradually moves to the right. Eventually the end of the specimen which contains the impurities can be cut off leaving a bar of pure germanium.

However, the germanium extracted from this oxide is very impure and as its use in transistors requires it in a pure form, careful refining is needed. This is done by a process known as zone refining. In this, heating-coils are slowly moved along a bar of impure germanium so that a short length of it becomes molten. Inpurities remain dissolved in the molten metal and, as the molten zone passes along the bar, the impurities are moved along with it. Eventually the impurities are concentrated at one end of the bar; this piece can be cut off leaving a bar of pure metal (see Figure 34).

When one considers the vast number of applications to which the transistor is applied (ranging from your radio or TV set to space satellites) it is difficult to believe that this device was only invented about twenty-five years ago. As recently as 1956 a Nobel prize in physics was given for work on semi-conductors and the development of the transistor.

Gold. This is one of the few metals which is found in nature uncombined with other elements. You may be surprised to learn that it is found widely throughout the World but unfortunately only a few of these deposits contain enough gold to make mining them worthwhile. Even in the richest gold deposits (mainly in South Africa but USSR is now an important producer) a tonne of ore may yield only ten grammes of gold. Thus the main problem in obtaining gold from the Earth lies in dealing with the vast quantities of waste materials.

The earliest method of obtaining gold was by 'panning' which involved washing away unwanted sand or gravel with a stream of water leaving the much denser gold. Today the same principle is still applied to hydraulic mining where powerful jets of water are directed on piles of gold-bearing gravel.

However, as gold particles are often too fine to to be obtained in this way, the rock is crushed and ground to a powder. This is treated with highly poisonous sodium cyanide solution which dissolves the gold leaving the rock. The gold can then be forced out (or displaced) from the solution by adding powdered zinc. The fine particles of gold left can then be melted and run into moulds, though the metal usually requires further purification.

Fig. 35 A general view of the West Driefontein gold mine in South Africa.

The bulk of the World's gold is not, as you might expect, used in jewellery but is hoarded away by Governments and banks. Although gold reserves are no longer as valuable an influence on World finance as was once the case, they are still extremely important. Incidentally, the term carat, (as in 22-carat gold) used in jewellery, means that the gold is really an alloy of 22 parts of gold combined with 2 parts of other metals such as copper or silver to harden it; pure gold is referred to as 24-carat gold.

Lead. The production and uses of lead have been described in the chapter on zinc since it is frequently found in mixed zinc/lead ores. The most common minerals of lead alone are galena in which the metal is combined with sulphur and cerussite (containing lead carbonate).

Magnesium. You are probably most familiar with magnesium as the result of your experience in the laboratory of burning the magnesium ribbon. In practice, magnesium only burns readily in this form or as a powder and machine parts made of magnesium alloys are difficult to ignite, though the fine turnings produced during machining may catch fire more easily.

The metal is too reactive to be found uncombined but is widely distributed in the Earth's crust as the carbonate (in dolomite and magnesite) and as the chloride (in sea water and the mineral carnallite). Perhaps the best known magnesium compound is the sulphate (called Epsom Salts because of its discovery in Epsom spring water) but this is not an important mineral.

Magnesium can be obtained by the electrolysis of molten magnesium chloride or by the reduction of magnesium oxide with carbon or ferrosilicon (an alloy of iron and silicon). High temperatures are required for this process.

Its low density (only 1.74 g cm³) makes it the lightest commercially used metal and its largest

Fig. 36 One of the main sources of magnesium and its compounds in the U.K. is the sea. These settling and reaction tanks are part of the works at Hartlepool in the North-East of England.

use is in light alloys with aluminium, for example in aircraft. These alloys are very easily machined but are readily corroded by acids and solutions of chlorides so usually require protection, particularly from sea water.

The use of magnesium metal as a sacrifice metal for protecting iron and steel from rusting has been described earlier. The metal is also used in flares and fireworks.

You may well be familiar with a suspension of magnesium oxide or hydroxide in water known as milk of magnesia which is often used as a remedy for stomach disorders. Magnesium oxide is very resistant to heat and so is frequently used for making furnace linings.

Manganese. This is found widely, usually in the form of oxides particularly as pyrolusite. The metal can be extracted by heating the oxide with carbon or aluminium as described for some other metals.

Manganese has little value on its own but is used a great deal in alloys particularly in steels. These are made much tougher, some typical uses being in railway points, crushing machinery and caterpillar tracks for tractors and tanks. Manganese oxide is used in dry batteries and potassium permanganate is a good disinfectant and antiseptic.

Mercury. This has long been regarded as a most unusual metal because of the fact that it is the only metal which is liquid at normal temperatures; it was known to the alchemists of old as quicksilver. In practice, when cooled sufficiently to solidify it, it behaves much as any other metal. Although it is occasionally found as the metal, its main ore is the sulphide, cinnabar, from which it can be obtained by heating this substance in air, the mercury going off as a vapour which can then be condensed. These vapours and also soluble compounds of mercury, are extremely poisonous.

The fact that mercury remains a liquid over a temperature range of $-39°C$ to $357°C$ and expands highly have led to its use in thermometers. It is also used for silvering mirrors, in mercury vapour lamps, and for making alloys known as amalgams. Amalgams with silver, copper and tin are used in dental fillings. An important use of the metal is as the cathode in certain cells for making sodium hydroxide and chlorine by electrolyzing salt solution.

Its compounds are used in medicine and for killing pests. An important mercury compound is mercury cyanate (perhaps better known as fulminate of mercury) used as a detonator for explosives.

Nickel. As we have seen, the main sources of nickel are the Sudbury ores of Ontario in Canada where its compounds are found mixed with those of copper, cobalt, iron and many other metals.

The metal is obtained as explained on page 29 and is purified by warming the impure metal in a stream of carbon monoxide gas. The nickel is converted to a vapour known as nickel carbonyl. When this is passed over nickel pellets maintained at a higher temperature it decomposes to produce nickel (which is deposited on the pellets) and carbon monoxide which can be used again.

Its main uses are in such alloys as nickel steels (some of which being tough, are used in armaments, while others are stainless) and cupronickel is used for coinage. It is widely used as the basic layer on steel before chromium plating and forms the base for many catalysts for speeding up industrial chemical processes such as the hardening of oils in the production of margarine.

Platinum metals. These are a group of metals which include platinum, palladium, iridium, osmium, rhodium, and ruthenium amongst others, of which platinum is the most important member.

Fig. 37 The processing of metals of the platinum type, although carried out on a small scale, is extremely profitable because of the high value of these precious metals. In this vessel, a mixture is treated with acids to dissolve platinum and palladium so that the remaining substances can be filtered off.

Although platinum can be found as the metal, sometimes naturally alloyed with some or all of the above metals, it is usually found in very tiny traces in other ores. The Sudbury ores for example contain minute quantities; although these amount to perhaps one two-thousandth of one per cent of the original mineral, the huge quantities of ores mined, coupled with the high price of platinum, makes its extraction worthwhile. The methods used are too complex to describe here but the processes are extremely careful and painstaking compared with those production methods already described. The high value of these metals makes it essential to reduce working losses to a minimum.

Most of the platinum metals are now used as catalysts in chemical production, perhaps the best known examples being the production of nitric acid from ammonia gas and the treatment of petroleum products to obtain higher grade petrols, the so called platinum re-forming process. Alloys are used for lining crucibles, for making certain high quality optical glasses, in thermo-couples (for measuring high temperatures) and in jewellery.

Plutonium. This metal was only discovered in 1940 and differs from all the others described in this book in that it was first discovered as a man-made element although it is now known that there are very tiny quantities present in the Earth's crust. It was soon realized that the nucleus of the plutonium atom was capable of undergoing fission (or splitting) with resultant production of large amounts of energy. A huge programme for its production was put under way in the USA with a view to use in a nuclear or atomic bomb.

Today it is of great use as a fuel for nuclear reactors, in electricity power stations and, in the so-called breeder reactors, more plutonium is produced during the process than is actually used up. A nuclear reactor can produce a million times more heat than could be obtained from the same mass of traditional fuel such as coal. Since World supplies of petroleum are being used up rapidly it may well be that in the future far more of our power will be obtained in this way though very great care has to be taken to avoid the atmosphere becoming dangerously polluted.

Silver. Silver can be obtained from sulphide ores such as argentite or as a by-product of the extraction of other metals such as copper, nickel, lead, zinc and gold. One method of obtaining the metal is by treating the ore with sodium cyanide; this dissolves the silver which can then be displaced by treating the solution with powdered zinc as described above in connection with gold. Another process uses electrolysis.

Apart from its decorative uses, silver is sometimes used as a lining in chemical and food-making plants (because it is unaffected by most substances) and as electrical contacts (because of its high conductivity).

The most important use of silver is in the production of silver nitrate, much of which is used in the preparation of silver compounds for photography. When these are exposed to light they become changed to a form which is easily reduced to silver by the use of suitable developers. The unchanged silver compounds (in the areas untouched by the light) can be dissolved away by solutions known as fixers. Thus on the familiar black and white negative, the black areas are in fact deposits of metallic silver.

Sodium. Sodium compounds are found widely and in another book of this series we shall be describing some of the many valuable compounds which can be obtained from sodium chloride or common salt. Sodium metal itself is produced by passing electricity through a molten mixture of sodium and calcium chlorides. It is a very reactive metal and has to be kept under oil to prevent it from combining with moist air. Much of the sodium produced is changed into a sodium-lead alloy which is used in the preparation of tetraethyl lead (an ingredient of modern petrols). In the future, petrol will contain less of this substance because it seems probable that the lead compounds thrown out from car exhausts are harmful; it seems unlikely that the inclusion of the compound in petrol will be completely banned so presumably the demand for sodium will continue.

Since it is so reactive, it can be used to reduce compounds of less active metals. For example, titanium chloride reacts with sodium to produce titanium metal and sodium chloride. It is also used as a conductor of heat in nuclear power stations.

Compounds of sodium are used in large quantities because they are usually very cheap and nearly all dissolve readily in water.

Tin. Tin is found in nature mainly as the oxide, tinstone or cassiterite. The main producer is Malaysia. This oxide is readily reduced to the metal by heating it in a furnace with coke or coal though the product needs further purifying.

Large quantities of the metal are used for tinplating steel for the canning industry. This is done by dipping the plates in molten tin or more usually nowadays by depositing the tin on the steel electrolytically; this produces a more even layer. It is also widely used in alloys, for example in the bronzes (with copper) and solders (with lead).

Titanium. This metal occurs widely in several forms. Although it is the ninth most abundant element on the Earth and was first obtained over 150 years ago it has only been available commercially for about twenty-five years. The ore is converted to titanium chloride by heating with coke in a stream of chlorine and the chloride is then reduced to the metal by magnesium or sodium, as described above.

Titanium is a light metal, less than twice as dense as aluminium, but is as hard and strong as steel. Moreover it is resistant to very high temperatures and to corrosion. These properties

Fig. 38 Nowadays most tinplate is made by depositing tin on steel sheet by electrolysis. The large rolls of tin as shown in the bottom right-hand corner are fed into a series of tanks which stretch into the far distance.

Fig. 39 Titanium is an expensive metal but has many valuable properties. This block of titanium weighing one tonne is being machined before forging, Note the operator's protective screen.

make it suitable for use in high speed aircraft and in rockets and missiles. Its resistance to corrosion makes it suitable for use in chemical plant and storage tanks.

Unfortunately its high melting point (200°C above that of iron) means that it has to be worked at very high temperatures. Under these conditions it absorbs oxygen which affects its properties and so it has to be worked in a vacuum or in the presence of an unreactive gas such as argon.

By far the most widely used compound of titanium is the oxide which is a white powder which has now largely replaced poisonous lead compounds as a base for paints.

Tungsten. This is a very heavy metal being twice as dense as iron. It is obtained from the mineral wolframite by first fusing this with sodium carbonate. The product is then treated with acid and the resulting substance is heated with hydrogen, carbon or aluminium to form the metal.

Its most important uses are in the filaments of electric light bulbs or alloyed with iron to make tungsten steels suitable for use as cutting tools.

Uranium

Another very dense metal, uranium, although distributed widely in nature, only became really important when nuclear fission on a commercial scale made use of the metal. Extraction of the metal is long and complex so that uranium is an expensive metal.

The need to conserve

As we have seen, metals play such a common part in our life that we tend to take them for granted. The deposits of minerals and the mines from which the ores are obtained seem so huge that we assume that they can be used for ever. Certainly, up to now, whenever important sources of metal ores have been exhausted, geologists have always been able to find a replacement, but this cannot go on for ever. Consequently, it is just as important for us to conserve our sources of metals as it is for us to make the most of our petroleum deposits.

We have seen that already some metals such as iron are recovered from scrap on a large scale and we know that precious metals are carefully looked after because of their high cost. It is now important that we take even greater steps to conserve our metals; if we go on using them at their present rate you may well find that within your lifetime some of the metals on which we depend so much may be no longer available and suitable replacements may not exist. We cannot always assume that new deposits of ores and new methods of production will continue to solve our problems.

Suggested activities

1 List the main methods of extracting metals which have been described in this book. Under each heading write down as many metals as you can which are obtained by each method. Compare this list with the electrochemical series in Fig. 10.

2 By consulting other reference books try to find out the names of as many metals as you can which have not been mentioned in this book. In each case try to find at least one important use. In some cases this may not be possible. Why do you think this is?

3 In producing many articles, plastics have replaced metals. Can you think of any examples of this? Why do you think that the use of plastics has been preferred in each of the cases you select? Metals are however still widely used in many cases as you have read. Why do you think they have not been replaced by plastics in these cases?

4 Find the names of three alloys which have not been mentioned in this book. In each case find the elements they contain and say why you think the alloys have been chosen rather than the pure elements?

5 What metal (or alloy) would you choose if you wanted to make each of the following articles: (a) a large statue to be mounted out-of-doors (b) a bridge (c) a small scale model car (d) the expanding part of a normal thermometer (e) as a seal in a fire sprinkler system so that the water would start flowing soon after the fire had begun. Explain how you made your choice in each case.

General projects and information available

Many companies producing metals and alloys have literature which is available free or cheaply, usually to teachers or adults though some may also supply limited materials to children who show a genuine interest. In this field there is also a number of trade organisations such as the Copper Development Association which produce material for a number of companies working in the same field.

The list is extensive and changes from time to time. The Schools Information Centre on the Chemical Industry, North London Polytechnic, Holloway Road, London N7 8DB (which has become a valuable source of information for schools) publishes at the present time a valuable information sheet listing many examples of such literature. These leaflets have the big advantage over a book in that they can be changed frequently to keep in touch with new material. Copies of this list are available free to schools at the time of writing.

Since iron and steel plays such an important part in this field, the British Steel Corporation's publicity department at 33 Grosvenor Place, London SW1 which produces a range of material should be given a special mention. A free list of publications is obtainable from the above address.